Vincent van der Meulen

# Building with a Positive Footprint

## Workbook



nai010
publishers

# Foreword

I grew up in Almelo, in Twente. In this Dutch region, concise formulation has been elevated to an art – no bullshit, no pretence, no hypocrisy, no deception, just tell it like it is. That is exactly what needs to happen now in the sustainability discussions in the construction industry. This book contains 19,000 words, and since the average person reads 250 words a minute, you can read the whole thing in less than two hours.

As an architect, I have been designing increasingly sustainable buildings for 15 years. But despite all the attention paid to the subject, the way we build in the Netherlands is still not sustainable enough. In fact, almost every building now being constructed around the world exacerbates our ecological problems. I am completely sick and tired of this situation, and besides, it's just not necessary. As a society we are not nearly ambitious enough, even though we already have the means to build so much better than we do. But that requires courage, looking beyond market conformity in our choices, intensive collaboration and a change in the way we collaborate in building projects. As an architect, I don't pretend to have all the answers; the only way we can do it is together with a great many partners. But the fact that you need others is no excuse for not doing what is necessary yourself first.

This compact book is meant as an accelerator, and the more people participate, the faster we can build in a radically sustainable way. I am sharing everything I know about working purposefully towards a new generation of buildings that actually do good. It gives you the insight to take steps yourself and keep everyone around you on their toes. We're going to build with a positive footprint.

# Reader's Guide

This is not a reading book – this is a workbook, so pick up a pen and write down your ideas immediately! I have left enough white space. My aim with this book is to get people in motion by answering four important questions. I begin with a common starting point: why do we build so badly at the moment? To answer this, I will briefly dive into history and describe where we want to go. I will then explain the methodology for designing a new generation of buildings that do good, that purify the environment. Afterwards, I will focus attention on the fact that this is not only better for ecology; we can also enjoy our buildings even more if we build with a positive footprint. Finally, I provide practical tips on what steps to take and how to take them in concrete terms, what you will come up against and how you can act effectively to deal with this. At the back of the book, there is a toolkit for the most important steps towards achieving a building with a positive footprint. This lets you get to work right away!

You can use this space to write notes – make this book your own! The dots help you to write or draw in a structured way.

# Contents

## Introduction

# Does Your Building Make the World a Better Place?

Is the world better off with your building? No is the answer in 99.9% of cases, at least for ecology. Even buildings with the highest achievable sustainability labels are generally still part of the problem, not part of the solution. We are living now in 2022, half a century after the report *The Limits to Growth* explained to us that there are limits to what ecology can bear. Yet with almost every building, we continue to increase the pressure on ecology. This must change, and quickly.

In the 15 years I have been working as an architect, I have also been part of the problem. I have seen how difficult it is to make a really radically sustainable building, because it can go wrong in so many areas. But I have also learned under what circumstances it does work. When it is possible to get everyone involved to contribute positively with the right intentions. When you know how to ensure that discussions remain no-nonsense – straightforward and to the point, and how difficult decisions are made with the right arguments. But most of all: how you find increasingly better solutions.

Unfortunately, this still happens far too little. Almost every building being constructed in the Netherlands today leaves a trail of destruction over the whole world in its wake, both during construction and in use. It can feel like a hopeless task to change this situation. It's true, it is difficult. We have made it too complex. That is why, in order to accelerate the transformation, we need to simplify the mindset, so that everyone can participate and contribute, generating understanding and enjoyment in the team. It means we can do the job well, instead of less badly. And make buildings that purify the air, instead of just polluting a little less. Buildings that generate energy instead of just using less.

Designing and building to do good, rather than less bad. To actively reduce, rather than increase, the pressure on our ecological systems. Making buildings that contribute positively. Building with a positive footprint. That is what I want to achieve with this book. Everyone involved in the building process must realize that technically almost everything is already possible, but that it is a question of choosing the ambitions, priorities and good partners. This makes everyone in the building process jointly responsible: architect, contractor, product supplier, consultants, users and, above all, the client. Everyone has an important role and everyone can contribute to making this mission a success.

The mission is important. Nothing less than the lives and happiness of our children and grandchildren are at stake, so we need to quickly learn how to design and build in a way that does not further jeopardize their future happiness. That is why the chapters in the book are related to Maslow's pyramid. First, we create the context. Then we cover how to get the base of the pyramid in order: with clean water, clean air, clean food and a safe, sustainable environment (physical needs and safety). We then show how buildings with a positive footprint are more meaningful and how we can enjoy them together (appreciation and social contact). Finally, we focus on how you as a person can contribute to buildings with a positive footprint (self-actualization).

I wrote this book to share my insights, method and lessons in an accessible way. Accessibility is crucial in order to give professionals and everyone involved in construction a common basis of understanding: experts, clients, users, architects, structural engineers and suppliers. With the help of this book, they can collectively answer the most important question: *How are we going to build with a positive footprint?*

# 0
# Why Do We Build Badly?

## Past and Present

It has taken me years to understand how the balance of power within the construction industry has led to us building in such a polluting way, and how design and construction have become increasingly disconnected from the nature and ecology surrounding us. This short history provides a common ground to further build on in the move towards a positive footprint. We need to understand the building culture in order to change it. I try to describe briefly and evocatively why we build the way we do and how history is now landing at the design team tables of our construction projects.

Imagine that you lived 300 years ago in, say, Gouda or Zwolle. After years of hard work and saving, you are going to build a house, so you approach a builder-architect nearby or helpful relatives. The house is built with the wisdom gained locally over the centuries; in Gouda, for example, you build on wooden piles. The bricks and other building materials are sourced locally, which make them affordable. They are of long-lasting quality. Energy is expensive, so the windows are not too big to avoid the cold in winter, but they should not be too small either, in order to get enough daylight in (candles are also expensive). Of course, the windows can be opened for ventilation. You collect rainwater from the roof, because it is free. The garden is used for growing vegetables and fruit, and the leftovers go to the animals or the compost heap. The entire construction chain involves perhaps a hundred people, most of whom you know. Maybe you buy a single piece of natural stone (all the way from Belgium!) as decoration for a family crest in the façade, as this special material from far away gives the house some allure.

In short, construction in the past was organized locally and largely sustainably, due to the high costs and scarcity of energy, water and materials, and the buildings were cleverly designed to function optimally in the local climate. Because the buildings in a town or city were often built by the same people, with the same materials and following the same logic, they form a common language. The houses are all slightly different, but belong to the same building culture with a local character, due to the materials and knowledge locally available.

And then you have the current construction industry. Usually, people buy or rent a house from a developer who has already built it to conform with market standards. The builder and architect won a tender on price and experience, and the architect would like to create something innovative. The materials are chosen on the basis of what the architect and users now find 'attractive' and then the cheapest materials that meet the specifications are sourced. Transport is cheap so materials often come from the other side of the world. The position and size of windows is determined by aesthetics and the views the windows can offer. The building services consultant recommends keeping them closed as much as possible for energy efficiency. Gas and electricity come in underground and are sourced far away. The garden is partially equipped with low-maintenance artificial grass and PVC garden chairs. Rainwater goes into the sewer and the plants are watered using drinking water. The whole construction chain involves perhaps a thousand people from all over the world.

Is the above an exaggeration? No. Not even for simple houses. Local building materials, ecosystem damage and energy have become global systems and as a result, the pollution, transport and damage have largely disappeared from view. Natural solutions have become mainly technological: for light, water, food, ventilation and building materials. The construction process has gone from being local and simple to global and fragmented.

Architects, builders and ultimately the residents do not see the damage when they flush paint down the toilet, select building materials from China, install an air conditioner or select a new PVC floor and get rid of the laminate floor.

And what happens when people do not directly see or feel something? It means they do not feel responsible for the consequences of their actions.

55

# Heads in the Sand

The last 75 years have shaped how we design and construct our buildings today: increasingly globalized, and increasingly as an economic activity for which the ecological health of the planet must make way.

After the Second World War, in the 1950s and 1960s, an explosion of prosperity and population growth began. Energy was cheap, technology was developing rapidly and population growth added to the economic flame. Construction was fast and low quality because of the scarcity of materials and money, and speed was more important than quality. Everywhere there was optimism!

But in the early 1970s came an oil crisis – bad for the economy, but good for the world as an ecosystem. We became aware of our energy wastage and our dependence on cheap fossil fuels. We started insulating houses, making more efficient cars, and researching sustainable technologies. In 1972, the Club of Rome published *The Limits to Growth*. In 1979, President Carter even installed solar panels on the roof of the White House in Washington.

In the 1980s, the oil crisis was almost forgotten. Exxon and Shell discovered the link between fossil fuels, global warming, rising sea levels and the global damage to ecosystems that would result. They thought that the problem should be solved mainly by governments, but to be on the safe side, they kept the reports secret. Reagan took the solar panels off the roof of the White House.

In the early 1990s, the Soviet Union collapsed and the raw capitalism that Thatcher and Reagan had started in the 1980s brought about strong economic growth. Morals were no longer expected of companies; they just had to make a profit, provided they operated within the rules. Companies were given a helping hand by deregulation and its twin brother self-regulation. The responsibility for choosing healthy or sustainable products was placed entirely on consumers. Companies could produce junk and nobody thought there was anything wrong with that, because 'consumers bought it'. Besides, if they didn't make it, someone else would.

In the early 2000s, China also joined the World Trade Organization and world trade became more global than ever before. The world was flooded with even cheaper products that were produced in an even less sustainable way. We consider this normal as long as we do not see the damage. We are now living in a globalized economy where shrimp are caught off the Dutch coast, peeled in Morocco, then returned to be eaten in the Netherlands. On the way, they are treated with chemicals to prevent decay. In the supermarket, they lie next to Asian shrimps that are a third of the price. It is no different with building products. Countries with low wages and no strict environmental regulations are ideal for making cheap building products, which are often mined crudely, produced in a dirty way and transported over great distances.

In the early 2010s, evidence was accumulating and awareness was growing that we need to produce, live and build differently. But our system was too good at producing cheaply and too good at covering up the damage, and we had become addicted to it. In fact, our thought patterns and discussions often follow the same logic. We want things to be different now, but we just don't know how anymore. We have buried our heads in the sand for 40 years and we have to relearn how to simply do things right. Obama filled the roof of the White House with solar panels again.

Now the 2020s are beginning. The past 75 years have seen tremendous progress in the fight against disease and poverty. Prosperity has increased enormously. For the quality of air, water, soil and biodiversity, however, these were unprecedented disaster years. So while we were busy shaping the top of Maslow's pyramid, we were invisibly punching a huge hole in the basis of our lives. At first, this happened without us realizing it, but for the past five decades, we have continued, like addicts, against our better judgement. We are just burying our heads in the sand and it is seriously scary. Billions of animals and millions of people die every year because of the often-invisible damage we cause. No one can foresee how the consequences of pollution and climate change will develop, but we are all responsible, we can all contribute and we all have an important role to play.

So what now? We must not only stop causing damage, we must quickly start repairing it. Every day, the patterns that have become ingrained over the past 75 years are still visible at the design team tables where new buildings are designed. We must quickly reprogramme the destructive patterns so that we can design a new generation of buildings that help to solve our problems. What we need is a generation of buildings that purify the air, water and soil, enhance biodiversity, reuse materials endlessly and generate energy.

# The Information Fog

The complexity and fragmentation of the construction process leads to an impenetrable information fog that makes building good buildings difficult. Imagine a meeting in a room with a hundred construction experts: contractors, designers, users, municipal officials, ecologists, the client, the bank, the district water board, the energy company, the structural engineer, electricians and suppliers. The latter supply bricks, natural stone, plastic wall cladding, air conditioning, sun blinds, wooden floors, PVC floors, green roofs, bitumen roofs, flax-wool insulation, PIR insulation and so on. The task before them is very simple: to build a super-sustainable house for a family. It must be 200 $m^2$ and it may cost 400,000 euro. The moderator begins: Let's start with the façade. What do we need to do to achieve a super-sustainable home? The room comes alight. A brick façade, produced locally, lasts forever! A wooden façade with FSC label! No, you have to paint wood, so then it is no longer sustainable! A plastic façade does not need painting and is maintenance-free for a very long time! Yes, but it is made from oil and will later become chemical waste! The meeting went on until late without a good result being achieved.

This is the information fog of the current construction process: too many people involved, too many definitions, too concerned with making money. Hardly anyone is just doing it right, and this is often partly due to the way the task is described. The starting points are not clear enough (what does sustainable mean?) or too specific (why 200 $m^2$?). The priorities are not clear (could we also build a smaller and more sustainable home?). But the authorities, suppliers and contractors involved also cause confusion. The parties contradict each other, lie and deceive to sell, and avoid taking responsibility. If problems arise after construction, everyone points the finger at each other. Recognizable?

There are so many parties involved, but nobody really feels responsible. Everyone can hide behind their own definition of sustainability, cost efficiency, incomprehensible reports, 'I-meet-the-legal-standards attitude' (the standards set as low as possible by the construction lobbies in The Hague and Brussels), market forces, and so on.

This information fog has a paralyzing effect. We have been talking about sustainability for decades, but in the meantime, almost every building constructed in the Netherlands is still a net polluter. Let that sink in for a moment: every single building that is built, including the sustainable ones, pollute the world even more. While technically speaking that's no longer necessary. So why is this still happening and where do we start with the solution? Of course, much stricter and more ambitious regulations would be the best solution – but that will take far too long. So we have to do it ourselves, without help from the authorities. And if we make that work, it feels even better.

To cut through the information fog, we first give everyone involved the same common understanding with this book. The same vocabulary with which to conduct the discussion. We then create a common goal by clearly formulating ambitions, starting points and priorities and ensure that everyone sees how they can contribute to achieving the ambitions in their own role. In addition, a good design process requires leadership – someone who keeps the process on track in ethical and conceptual terms. Preferably the client or user, but it could also be the architect. That person leads the process and keeps the conversation going in the right direction: cutting through the information fog to move towards a building with a positive footprint.

# Don't shit where you eat

I once dreamed of a house. Just an average house in an average Dutch neighbourhood, but there was something special about it. All the damage and pollution that had resulted from the construction of this home had been collected inside. Everything. Through the window, I saw the living room completely filled with construction debris. I went inside, but I could not breathe there. All the diesel, kerosene and coal vapours had been captured and pumped into the house. All that dirty air was released in the transportation of the materials from all over the world, by the bulldozers, and in the energy generation for the factories. The kitchen was full of dead animals, plants and trees that had been wiped out during the extraction of the raw materials. I walked upstairs. The bathroom and toilets were filled to the brim with heavily polluted water full of dead fish – the water that had been released in the production of the paint, coatings and building products. The bedrooms were filled to the ceiling with packaging from building products, polystyrene foam, cardboard, plastic film, empty paint pots, brushes and broken pallets. The residents were standing outside, because there was no room left in the house to actually live there.

This house, of course, symbolizes every office, every shop, every school, every town, and therefore the entire built world. There are two fundamental problems that we need to address in order to build with a positive footprint: the blinkers that prevent us from fully understanding the situation, and our short-term thinking. Don't get me wrong – I truly believe that most people and companies have good intentions. But we are putting our priorities and our focus in the wrong place.

First, the short term. We want to satisfy short-term needs, producers and investors want to make money in the short term, and we want the sustainability label that is currently in demand. When it comes to finances, we mainly look at the costs (in the short term), forgetting that investments can have a positive impact during the period of use by the occupants, and the next occupants, and the occupants after that. We need to think more in the long term, because that is where you often find major gains that we miss due to our short-term thinking.

Then our blinkers. The damage we do mostly takes place outside our view. The major changes in the climate and the decline in biodiversity are happening so slowly that we hardly notice them. We rarely notice the air pollution; we burn our waste or hide it under a 'lovely' green hill; the impoverished soil quality is something only farmers, plants and animals notice. We don't see it, so it doesn't exist. We have deliberately put on our blinkers to avoid seeing that we shit where we eat.

But if you do see it and know it, you can also do something about it. That can be incredibly satisfying. Not just for you, but for the entire team you work with. Remember for a moment that house full of rubbish. It would be amazing for the residents if they could walk into a clean house, wouldn't it? What's the lesson here? Don't shit where you eat!

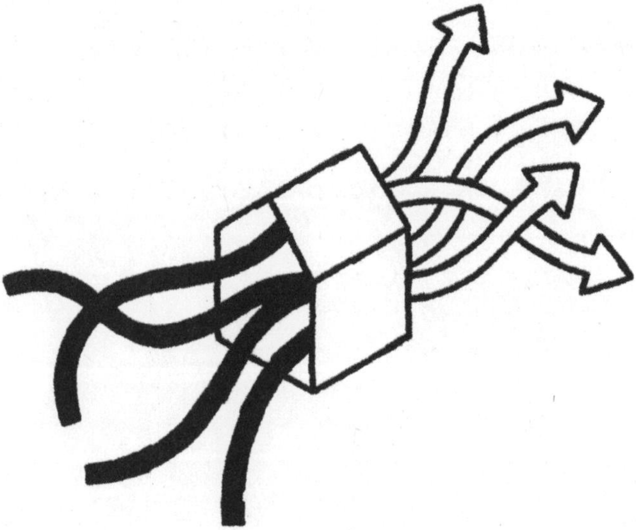

## Design to Do Good

So, we must be more ambitious, and concentrate on a positive footprint. Now imagine a large kitchen table with ten driven construction experts. Almost all of them know each other from a previous project and they now have to build a radically sustainable house. The clients want something very simple: a house with a positive footprint on the world – a building that does good. They know what that means. It is a house that purifies the air and water; that generates more energy than it uses; where after completion there are more animals and plants alive than before; that is built entirely from clean materials that can be 100% reused; that makes the surrounding soil more fertile while it is being used; and where a sustainable, healthy lifestyle is made easy for the occupants. Why should you settle for less? Especially when you are building a new house that may still be standing two hundred years from now.

The architect starts the conversation. The ambitions and priorities are fixed – we will not settle for less! The future residents are also at the table; they want to be able to live there comfortably with the four of them. They have about 400,000 euro for the construction costs. What do we need to know about them to make their dream home? The table comes to life: do they have hobbies that we should take into account? Do they like gardening? I recently made a very beautiful butterfly garden on the roof of a building. I have found a wood-construction system that does not contain any glue. Perhaps we can purify water via the pond. The meeting went on until late, with amazing results!

It has become normal to construct buildings that exactly meet the current economic, fiscal, technical and moral reality of market conformity. This leads to buildings that are actually a minimum viable product – not a mature, developed product. Something has been created as quickly and cheaply as possible that meets the minimum requirements, and which people are willing to pay for. In these minimum requirements, sustainability is at the bottom. In the design process, the ambitions are often higher. But regulations, the functional and financial requirements and the planning are always written in stone (the project managers and developers are even proud to point this out), so that with every setback, the sustainability ambitions are adjusted further downwards. Regulations such as the Building Code and the legal energy requirements represent the lowest possible bar. Anything in excess of that is seen as a waste of money.

What happens if we turn it around and stop going for less bad, but start designing to do good? We make sure we have a positive footprint, *that* is the bottom line, while the finances, size and functionality become variables. Every time I suggest this, all the building professionals look at me as if they cannot believe what they are hearing. But is it really so very radical? No, it is actually radical to design something cheap that makes only you happy and then dump your rubbish in someone else's garden. All I am proposing is that we design a new generation of buildings that do good in ecological terms. 'Easier said than done!' 'Where on earth do you start?'

**Make it simple:**

1 > **See what flows go into a building, and make sure they come out cleaner.** You should be able to explain it to an eight-year-old.

2 > **If you can't do it within the building boundaries, overcompensate.** If you cut down one tree, then put two or three back!

3 > **Always communicate about this with radical honesty and share the lessons.** No slick marketing, but clear information you can learn from.

**Technically this has been possible for a long time.** We just have to want it.

# I
# How Do We Design Purifying Buildings?

## 7 × 7 = Positive Footprint

Air
Water
Energy
Biodiversity
Soil
Building Materials
Food

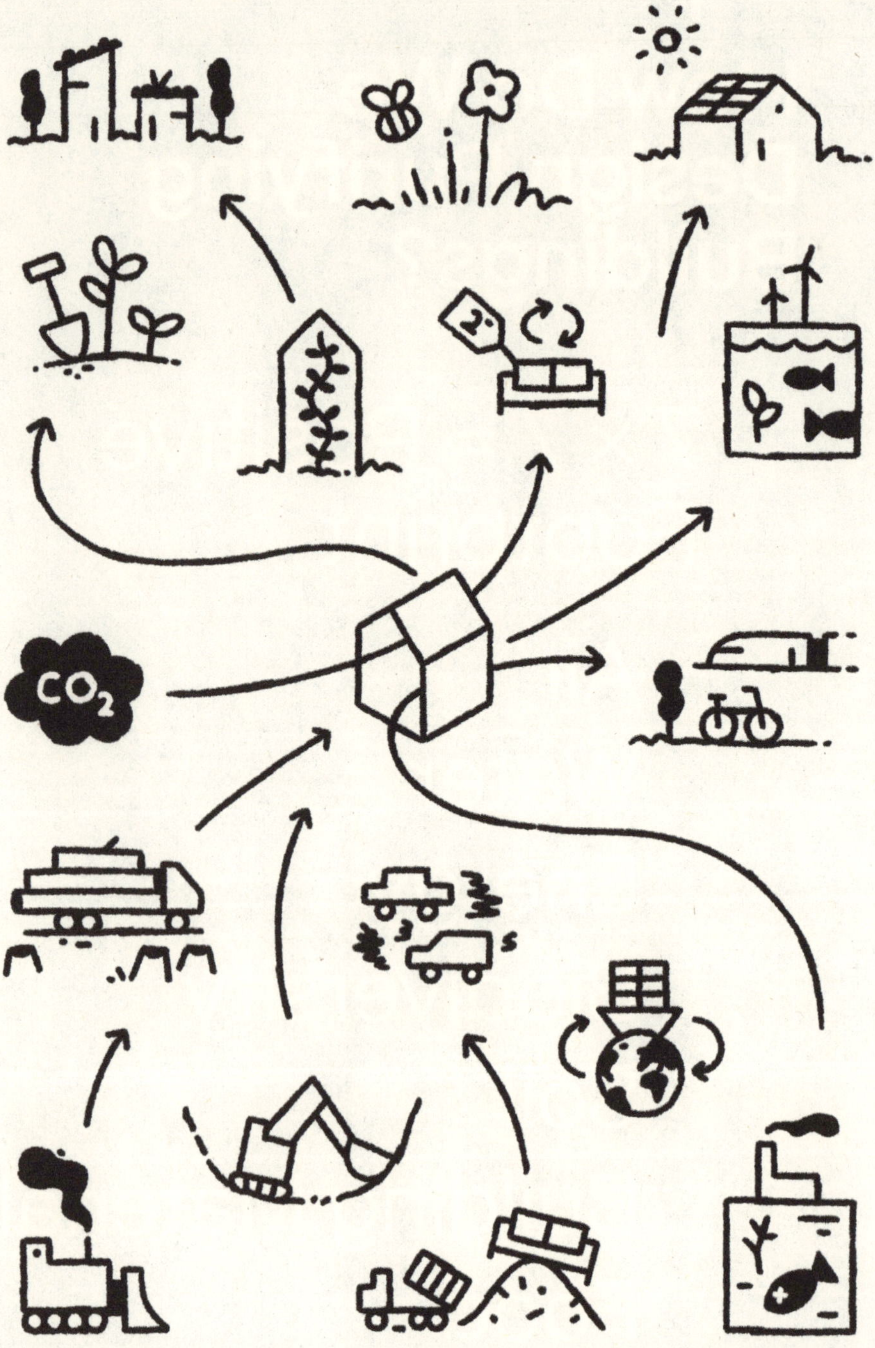
CO2

# 7 × 7 = Positive Footprint

This will sound ridiculous to some people. But the next bit is perhaps the most radical proposition of this book:

**Determine what goes into the building, and make sure it comes out cleaner.**

That is the definition of a building with a positive footprint. That is all you need to do. It is of course easier said than done. Because a great deal goes in and out of a building during its lifetime. But we are going to get there step by step. A building is a place where different flows come together. I distinguish seven of them: air, water, energy, biodiversity, soil, building materials and (waste=)food. Together, they form the basis that fulfils our physical needs and provide safety. Clean water, food, air and shelter: these are the basics we need to get right first.
For all seven flows, the approach to achieving a positive footprint is the same. It consists of seven steps. 7 × 7 = a positive footprint.

## 1 > Formulate the Positive Ambition

Make sure that this ambition is simple and understandable for everyone who wants to contribute. It may be very general, for instance: the building purifies the air. Or something more specific: the building purifies the motorway air to become forest-air quality. Make sure that your ambition is achievable, and has a net positive effect on the world, but does not include the specific direction towards a solution.

## 2 > Measure the Baseline Situation

In relation to what do we want to have a positive footprint? The question we want to answer here is, for example: what is the quality of the air we are drawing in now? How much biodiversity is there at the location now? What has surprised me most in my work is that this question is not asked. And no one has the knowledge at hand. Not even an indication. Ask for these 'baseline measurements' at an early stage.

## 3 > Harvest-focused

You know what you need to fulfil your ambition and to end in a positive position compared to the baseline situation. Now, for example, you have to harvest your energy, water and air in a smart way so that it can be used by the people. Some ways to do this are to harvest sunlight for electricity or building materials from another building. What is important is that you harvest in such a way that you can achieve a net positive effect. If you bring in dirty air, for instance, you can filter it and make sure it comes out much cleaner.

## 4 > Use and Enjoy Wisely

You have a clean flow. Enjoy it! And make sure that people use the flows wisely, so that the system does not become unnecessarily polluted or over-used. Keep it simple and make sure people understand how they can influence their environment themselves. For instance, by using water-saving toilets, windows that open and biodegradable cleaning products.

## 5 > Store What is Needed

Not all flows come in at the moment they are needed. So, save the flows for later use. In the same way, for example, that we can store heat in the summer and use it in the winter.

## 6 > Purify Afterwards, Out of Love

After use, you send the flow back into the environment. You purify it after use, so that it goes out cleaner than it came in. If you borrow clothes from someone, you wash them before you give them back, don't you? For example, you purify toilet water before it can go back out again. Sometimes this step also means that you have to compensate, or overcompensate, because you cannot solve something at your location. For example, $CO_2$ emissions released during the production or transport of a particular material.

## 7 > Measure and Communicate

Naturally, we want to be absolutely certain that we have a positive footprint and that is why we measure how clean each flow is as it leaves the building. This ensures that we can continue to improve the systems and we can also proudly communicate about them (again, just in a no-nonsense way please, no misleading marketing).

*Voilà, c'est tout.* 7 flows × 7 steps equals a positive footprint. The tangible solutions can take many forms and are unique to each building. In the continuation of this chapter, I will explore each flow and step in more detail. You can use the matrix from the toolkit to get a handle on the flows and steps.

# Air
# More than just $CO_2$

It is clear to everyone how important clean air is for your health, concentration, sleep and so on. Air pollution comes from outside (cars, planes, factories) and inside (paints, coatings, adhesives, air fresheners). People benefit from plenty of ventilation, but it is better for energy consumption to have minimal ventilation. Each building must find its own balance, but I would like to give two rules of thumb. Firstly, always make windows that open. Secondly, when faced with a difficult decision, always choose better air quality for people and animals. Any extra energy that this requires can be generated sustainably.

# Air 7 Steps

## 2 > Measure the Baseline Situation

Make the assessment on the basis of occupancy and standards. Then add 20% because ventilation standards will become stricter in the future. Measure the air quality at the location. At the very least, measure the levels of $CO_2$, CO, $NO_x$ and the different types of particulate matter in ppm.

## 3 > Harvest-focused

Draw in air where it is as dirty as possible. That way you can clean more. Use landscaping with plants, trees and olivine to purify beforehand. In the building, you use technical filters to increase the air quality. Ask for additional filtering compared to the standard. After that, the air can flow into the rooms.

## 1 > Formulate the Positive Ambition

The building purifies the air.

## 6 > Purify Afterwards, Out of Love

Choose materials that have a positive effect and, for example, bind dust and particulate matter, or plants that bind particulate matter and produce oxygen. The air leaving the building can be further filtered by plants in the roof or façade and/or by technical filtration. You can also choose to ensure that you offset the $CO_2$ production of the building's users. By planting trees, for example. This requires a lot of space and is better done off-site (in combination with compensation for material use).

## 7 > Measure and Communicate

Measure the quality of the air entering and leaving the building. Communicate the difference and let everyone know. This building purifies the air! Also check whether building users feel the difference.

## 4 > Use and Enjoy wisely

Design the building so that windows can be opened. Preferably in such a way that the building can be naturally ventilated for a good part of the year. Create green outdoor spaces so that people can get fresh air. Give people the power, but let them know if they are placing excessive demands on the system (heating or cooling switched on while the window is open). In the case of $CO_2$-controlled ventilation, make sure you set the threshold lower than the standard.

## 5 > Store What is Needed

When ventilation is not taking place (from an energy point of view), the air is stored. But it must not be contaminated by the materials we have inside. So, for example, make sure that the materials and cleaning products you choose do not release any volatile organic compounds (formaldehyde, terpenes, flame retardants and plasticizers). High ceilings also provide a 'buffer' of fresh air, making the air in a room feel fresher for longer.

# Air Worksheet

### 2 > Measure the Baseline Situation

### 3 > Harvest-focused

### 1 > Formulate the Positive Ambition

6 › Purify Afterwards, Out of Love

7 › Measure and Communicate

5 › Store What is Needed

4 › Use and Enjoy wisely

positivefootprint.nl/toolkit

# Water
# The source of life

Your body is 70% water. Water is also the flow we can enjoy the most. When we are thirsty, looking at the sun glitter on the water, when we play sports in it or on it. There are enough reasons to take this flow seriously and certainly in 'Nederland Waterland' we should set a good example when it comes to water. In every single building.

## 2 › Measure the Baseline Situation

Make the assessment of how much water you will use, based on the occupancy and standards. Then add 20%, as usage may vary. Measure the water quality at the location of the rainwater, groundwater and surface water. At the very least, measure the levels of acidity, ammonium, nitrate, phosphate and coli-group bacteria. Also estimate how much rainwater you can collect per year. Now you know approximately how much water you will use and can purify.

## 3 › Harvest-focused

There are three types of water flowing in: rainwater, drinking water and surface water. Rainwater and surface water must first be filtered technically or naturally before we can use it. In the Netherlands, it is forbidden to make your own drinking water, but you can purify it and remove the calcium, for example. Harvest each water stream for the use that suits it. For example, rainwater is fine for cleaning and irrigation, but not as drinking water.

## 1 › Formulate the Positive Ambition

The building purifies water.

## 6 > Purify Afterwards, Out of Love

This is the most difficult step. There are two flows, grey water and black water. In the Netherlands, you are obliged to have a sewer connection and it costs a lot of money to purify your own sewage water. Nevertheless, you should try to do so. By yourself, or together with those around you. You can extract valuable nutrients from the water. You can do it using technology, for example with a bioreactor, and biologically, for example with helophyte filters. Both methods ensure that the water can be discharged back into the surface water (it does have be cleaner). You have nutrients that can be used as a valuable residual flow for agriculture or biogas, for example.

## 7 > Measure and Communicate

Measure the quality of the water entering and leaving the building. Communicate the difference and let everyone enjoy the clean water.

## 4 > Use and Enjoy wisely

Water purification is an activity that consumes significant energy and/or surface area. It is therefore important to conserve as much water as possible. Water is valuable, so treat it as such. In use, the clean water becomes black water, grey water or evaporates. Keep the water as clean as possible while it's being used. For example, use biodegradable toilet paper and cleaning products. Or better still, hang a small shower head next to the toilet so you don't need paper. Just like billions of other people.

## 5 > Store What is Needed

Rainwater can be stored using a tank or drainage crates under green roofs or landscaping, for instance. Ensure that rainwater can infiltrate the ground for good soil quality. In addition, store excess rainwater and then use it in dosed quantities in toilets or for irrigating plants. Make a rainwater tap for this purpose. Temporary storage is also needed for the purification. Water usage comes in peaks, while purification works at a constant rate.

Toolkit

# Water Worksheet

## 2 > Measure the Baseline Situation

## 3 > Harvest-focused

## 1 > Formulate the Positive Ambition

4 > Use and Enjoy wisely

5 > Store What is Needed

6 > Purify Afterwards, Out of Love

7 > Measure and Communicate

positivefootprint.nl/toolkit

# Energy We know what we have to do, just do it!

This is the most talked-about flow, especially in relation to the greenhouse effect. Thanks to regulation, there are often profitable investments that can be made in energy generation or conservation. That sounds great, but in reality, an energy-positive building is rarely delivered. Particularly rare is a building that generates both use-related and building-related energy itself. Mostly due to unwillingness, but sometimes it is just not possible to generate enough energy at the location. Then you have to make a side step and generate your energy in a green way somewhere else. For the energy flow, the possible techniques are already widely known. It mainly comes down to setting the ambitions and implementing them.

## 2 > Measure the Baseline Situation

Estimate how many kWh of energy you can capture at the location with sun and wind. Assess how much energy you will need annually for your building's usage needs, including keeping it warm. Add a 20% margin to that and you know what your challenge is.

## 1 > Formulate the Positive Ambition

The building generates more energy than it uses.

## 3 > Harvest-focused

Focus your building design on the sun for energy and daylight. In addition, consider the seasons. For electricity, it is simple. There are currently only two meaningful ways: solar panels and windmills. Use them efficiently. Heat and cold can be extracted from the environment, from water, from the air or from the ground via heat pumps. Solar water heaters can provide hot water using the heat of the sun.

## 4 › Use and Enjoy wisely

It is easy to go overboard with energy-saving behaviour and lose the soul and enjoyment of your building. For example, creating buildings with no windows that can be opened. Beware of that, but do make sure you have presence detection for lighting, a low base temperature and a light level that people can upgrade themselves, heat recovery, and eco settings on computers and machines that go into hibernation mode. This all helps. The better you do this, the more clean energy you can supply to the world.

## 5 › Store What is Needed

Electricity is best given to the grid. When the grid is full, you can store it in a battery or charge your electric car, do your laundry and heat water. Store heat and cold underground, in PCMs or in the mass of the building.

## 6 › Purify Afterwards, Out of Love

Electricity does not need to be purified afterwards. You can, however, see if you can supply residual heat or cooling to your surroundings. In that case, it may be necessary to raise the temperature first. Do this yourself, sustainably.

## 7 › Measure and Communicate

Measure your energy-consumption patterns and try to anticipate them. Communicate the energy consumption and let everyone know what they can do to reduce it even further. In addition, make your positive footprint clear in an understandable way. For example: in addition to our own energy, we have also generated renewable energy for 12 households!

# Energy Worksheet

2 › Measure the Baseline Situation

3 › Harvest-focused

1 › Formulate the Positive Ambition

## 4 > Use and Enjoy wisely

## 5 > Store What is Needed

## 6 > Purify Afterwards, Out of Love

## 7 > Measure and Communicate

positivefootprint.nl/toolkit

# Soil
# The poor relation of all the flows

The soil is the most undervalued flow. No one really cares about it in the construction process, apart from the developer if he comes across contaminated ground and it becomes a cost item. But the poison we have pumped into the ground and the impoverishment of soil quality threaten our clean water, our food and biodiversity in general. The causes lie in agriculture, infrastructure and the built environment with its industry and pollution. So it is a severely underestimated problem.

## 2 > Measure the Baseline Situation

Measure the soil quality at the location. What is its composition? What does it contain in terms of life? What does it contain in terms of pollution? To have a positive footprint, it helps to build in places where there is already no longer any fertile soil. So, for example, on the site of an existing building, car park or industrial estate. Once you know what the contamination is, you can look for the plants or techniques to purify it. If the composition is known, you know what can grow on it; if you know the life present, you know how to support or enrich it.

## 3 > Harvest-focused

You need soil and you are probably going to move it. Purify the ground you touch; the technologies are getting cheaper and can take place on site. In addition, disturb the soil as little as possible. So, for example, keep your construction site compact to minimize disruption to the soil and the life it contains.

## 1 > Formulate the Positive Ambition

The building purifies the ground.

## 6 > Purify Afterwards, Out of Love

If you have tried everything and it is really not possible to improve the soil on your site compared to the situation before you started building (for example, if you are building in a meadow), then look for ways to compensate and compensate very generously. Preferably in a tangible way that can be experienced in the vicinity: for instance, turn a car park into a 'tiny forest'. If that is not possible, make sure you compensate with, for example, Justdiggit or another greening project that improves the soil.

## 7 > Measure and Communicate

Measure the quality of the soil every five years. Communicate the difference and let everyone take home clean compost.

## 4 > Use and Enjoy wisely

Make sure that the soil stays clean or becomes cleaner during use. Be alert to cleaning products that can get into the ground when washing windows, for example. Keep diesel and petrol cars away. Select plants that help the soil regain its strength, such as reed grasses, willows, cabbage varieties, poplars, yarrow, Indian brown mustard and foxglove.

## 5 > Store What is Needed

For contaminated ground, there are soil depots where we temporarily store the soil. But in and around your building, it should be about purifying and improving. This could be in the form of compost or natural soil improver. If you have properly separated your technical and biological cycles, a good compost system will allow you to continuously improve the ground and create healthy soil storage.

Toolkit

# Soil Worksheet

## 2 > Measure the Baseline Situation

## 1 > Formulate the Positive Ambition

## 3 > Harvest-focused

### 6 › Purify Afterwards, Out of Love

### 7 › Measure and Communicate

### 5 › Store What is Needed

### 4 › Use and Enjoy wisely

positivefootprint.nl/toolkit

# Biodiversity
# Our greatest challenge of all

People often poke fun at nesting boxes for birds and bats. Something for eco-freaks. But biodiversity is deadly serious. It is perhaps our biggest global problem. It is a prerequisite for us to live, just like clean air and water. The difference is that the consequences of biodiversity loss are more difficult to predict. You can purify polluted water, but if a species becomes extinct, it cannot be reversed. As soon as we really notice the consequences of biodiversity loss, it is already too late. That is why action is needed now. In every building, we can make a difference by providing animals and plants with food and shelter. And by preventing nuisance. After all, it's wonderful to see breeding birds from your office!

## 2 > Measure the Baseline Situation

Measure the biodiversity at the site. Involve a local ecologist. For example, use the system of the Netherlands Environmental Assessment Agency that works with nature points. Nature points = area × nature quality × weighting factor. For a positive footprint, it helps, as with soil quality, to build in places where there is little biodiversity. So, for example, on the site of an existing building, car park or industrial estate.

## 1 > Formulate the Positive Ambition

We create more biodiversity than was present on the building site before construction.

## 3 > Harvest-focused

Together with a local ecologist, ensure that your design gives the local ecosystem a boost. The ecologist knows which species are under threat. Find room for plants and animals in, around, on and against the building. Tell the structural engineer directly from the start to take into account a substrate layer for the green roofs of at least half a metre, preferably one metre. Be creative – even mosses and mushrooms can grow in underground car parks. Use trees on the east, south and west as sun blinds, combining biodiversity with functionality.

## 6 > Purify Afterwards, Out of Love

If you have tried everything and it is really not possible to improve the biodiversity compared to the situation before you started building (for example, if you cut down a wood), do not build! Or look for very generous compensation. Preferably something tangible that can be experienced in the vicinity. Create a huge green roof on a nearby industrial hall, or buy an empty industrial hall, dismantle it and plant a wood. If that doesn't work, make sure you compensate with, for example, Justdiggit or Treesforall. However, compensation is really a last resort.

## 7 > Measure and Communicate

Measure the quality of the biodiversity. Celebrate the frog spawn and eggs laid: there are more of them than last year! Communicate the progress and let everyone enjoy it.

## 5 > Store What is Needed

Storing biodiversity does not make much sense. But the plants and animals inside the buildings are a kind of stored biodiversity. You can use flowers and plants here, get creative! Indoor greenery contributes to acoustic comfort. It improves and humidifies the air. Water everything with rainwater.

## 4 > Use and Enjoy wisely

Also, make sure you design to be as maintenance-free as possible. This way, nature can take its course and find its own balance. Besides, it will cost you less. If you do want to irrigate, use rainwater. Select a diversity of plants that help the soil, purify the air, clean the water and contribute to biodiversity by, for example, providing food or shelter. Make sure people look out onto the beautiful greenery. That they can have lunch there. Involve people who have green hobbies or look after an animal together.

Toolkit

# Biodiversity Worksheet

## 2 > Measure the Baseline Situation

## 3 > Harvest-focused

## 1 > Formulate the Positive Ambition

## 4 > Use and Enjoy wisely

## 5 > Store What is Needed

## 6 > Purify Afterwards, Out of Love

## 7 > Measure and Communicate

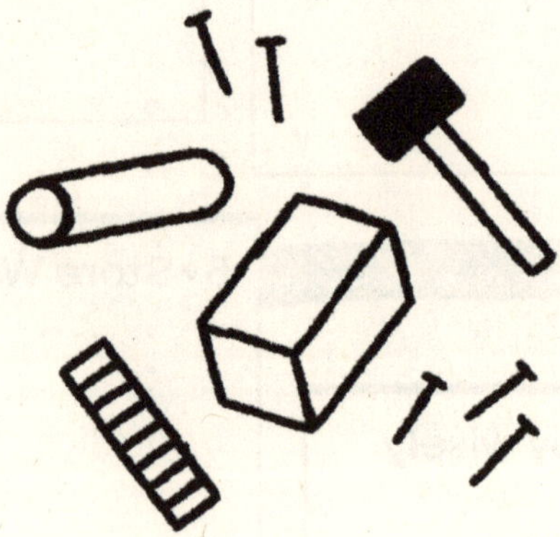

# Building Materials
# Would you let your children drink it?

The impact of materials is getting more and more attention, as we find out how wasteful and unhealthy our way of dealing with them is. A strategic scarcity of raw materials is also looming on the horizon. So, we have to learn to be smarter about how we use our materials. Which brings us to the circular economy. It sounds simple, but it is very difficult to execute well. Keep the technical and biological material cycles pure. And design it so that they can be reused endlessly.

# Building Materials 7 Steps

## 3 > Harvest-focused

Design for the long term. The longer materials do not have to be reused or recycled, the better. Preferably select recycled materials, and otherwise use cradle-to-cradle certified materials or biodegradable materials. Be extremely critical, also of the added protective coatings of materials, because suppliers will try to convince you of all sorts of things with their information fog. In the case of a difficult chemical formula, ask yourself: would you let your children drink it? Create a demountable design, so that materials can be easily harvested again when their technical lifespan is over. The purer and more demountable your material is, the more the demolisher/harvester will pay for it. The most difficult part to make circular are the machines that are in a building.

## 4 > Use and Enjoy wisely

Record all materials in a materials passport. That is the inventory of your raw materials warehouse. The warehouse will increase in value due to increasing scarcity. During the period of use, a lot can go wrong. A lick of paint over an untreated wooden beam? Chemical waste. Challenge the installation consultant early on in the process to examine how to avoid building-services technology. The reason is that many machines are full of plastics and hard-to-separate metals and chips. For all machines with many complex chemical materials, you demand a take-back guarantee from the manufacturer. Leasing lifts or lights, for example, is also an option that is in line with this idea.

## 2 > Measure the Baseline Situation

Use materials that have been used before as much as possible. For existing buildings, make a harvest map of which materials are available and can be reused (in a healthy way). In new construction, look for demolition buildings in the area to harvest materials. When selecting building materials, look first at chemical purity, and only then at CO2 emissions. Otherwise, any discussion will come to nothing. So, for example: pure concrete is better than laminated wood with chemical glue. Laminated wood held together by dowels or biodegradable glue is better than pure concrete.

## 1 > Formulate the Positive Ambition

The building will be a valuable source of clean materials in the future.

## 5 › Store What is Needed

Your whole building is one big raw materials depository. Prepare demolition specifications in the design phase. So that everyone knows not only how to make the building, but also how to dismantle it and exactly what it contains.

## 7 › Measure and Communicate

Your materials passport is the most important means of communication. Keep it well up to date, when rebuilding and renovating. Also communicate to people where distinctive materials come from and why they are good and pure. For example: our wood comes from Staatsbosbeheer (Forestry Commission) in the Dutch province of Flevoland. The concrete comes from a demolished furniture factory in Utrecht. The building consists of 30% recycled concrete by mass, 25% untreated and galvanized steel, and 15% C2C-certified glass.

## 6 › Purify Afterwards, Out of Love

The production, transport and use of building materials (concrete, in particular) releases $CO_2$ and other harmful substances such as $NO_x$, $SO_x$ and various types of particulate matter. Wood comes at the expense of forests, while textiles involve water pollution. This absolutely must be compensated in order to achieve a positive footprint. You calculate your ECI (Environmental Cost Indicator) with the available software (e.g. for MPG). This gives the impact in 11 categories. So, in the surcharges on building costs, there will be an item: compensation ECI building materials. Start with 10% of the construction costs. This is spent on forest planting, water purification or other compensation projects. If you design more smartly, choose more sustainable products and build using electricity, this compensation item will be lower and the amount can be invested in the quality of the building. If you sell or leave a building, make sure the materials passport and maintenance instructions are handed over properly!

Toolkit

# Building Materials Worksheet

3 > Harvest-focused

4 > Use and Enjoy wisely

2 > Measure the Baseline Situation

1 > Formulate the Positive Ambition

7 › Measure and Communicate

5 › Store What is Needed

6 › Purify Afterwards, Out of Love

positievefootprint.nl/toolkit.

# Food
# Stop calling waste waste

A building with a positive footprint should also be used with a positive footprint. In the usage phase, there are large flows of things running through the building. Things such as paper, food, furniture and electronics. These can also have a major environmental impact. There should be no waste; all flows should be food for the next life of the product or material. The cradle-to-cradle philosophy explains this perfectly. As long as we treat waste as something worthless, nothing will change. Treat it as food for the next life. Buy, use and sell valuable products that last and are easy to reuse or recycle.

## 3 > Harvest-focused

What comes in must come out. It all starts with sourcing from sustainable suppliers. This applies to everything used in the building: the tea bags, paper, printers, chairs, food, pens, cleaning materials and so on. Pay attention to the product itself, but also to the producer, the production method, the transport and the packaging. Services such as insurance, mobility and telephony can also be purchased sustainably. This is a lot of work once (think of it as moving house) and after that it is no longer a problem. A platform such as c2ccertified.org will help you get started practically and quickly. Choose sustainable alternatives that are also directly tastier or more fun. Like a tea strainer instead of a tea bag or a Tesla instead of an Opel.

## 2 > Measure the Baseline Situation

Make an assessment of the biological and technical flows going through the building. This gives insight into the scale of the recycling station and composting. Involve the people who do the purchasing directly. Without them, a positive footprint is impossible. You want to create pure biological cycles and pure technical cycles. If you buy products that cannot be separated, it goes wrong. For example, tea bags made of paper reinforced with plastic fibres.

## 1 > Formulate the Positive Ambition

We produce clean, valuable residual flows.

## 4 > Use and Enjoy wisely

Make it enjoyable for your people and make the sustainable choice the easiest. Prevent waste by using smart printing systems or a sustainable lunch provision, for example. Create attractive recycling stations divided into paper, glass, plastic, compostable and coffee. Preferably make the bins transparent. Make sure that small complex products, such as pens, electronic waste, batteries and light bulbs also have a permanent collection place here. In high-rise buildings, consider switching to an intelligent chute system to make it easier for the residents. It is also important to use your products for as long as possible. So be sparing with them. Especially your electronic products, because they have a huge environmental impact.

## 5 > Store What is Needed

The technical and biological flows are processed separately. The technical materials go to the container space. This has to be a pleasure to visit. Make it a bright, nicely finished room with a transparent door. Make sure there is plenty of ventilation, preferably by means of windows that can open. The technical materials are collected for reuse either by the local council or by other contract partners. The biological flow goes directly to the compost heap (on the roof or in the vicinity). Here it is processed into a soil improver for and by nature.

## 6 > Purify Afterwards, Out of Love

There are many inspiring alternatives to deploy your flows as valuable raw materials. For instance, companies that collect your coffee grounds to grow mushrooms on them. Become familiar with what the neighbourhood requires, so that your pure flows are reused directly and not mixed with less pure flows. You might even get fresh mushrooms in return!

## 7 > Measure and Communicate

Measure how many kilos of technical and biological flows leave the building each year for the next life. Limit this a little more each year, or make them more valuable. If you have established inspiring partnerships for your residual flows, communicate that proudly. It's a good way to inspire your team and others. Careful separation will increase!

Toolkit

# Food Worksheet

## 3 › Harvest-focused

## 2 › Measure the Baseline Situation

## 1 › Formulate the Positive Ambition

### 4 › Use and Enjoy wisely

### 7 › Measure and Communicate

### 6 › Purify Afterwards, Out of Love

### 5 › Store What is Needed

Mextropoli Pavilion, Mexico  Dellekamp + Schleich architects  Photo: Sandra Pereznieto

# II How Can We Enjoy Clean Buildings Together?

## Clean Flows Become Fantastic Places

Now you know the step-by-step plan for the base of the pyramid: the method for cleaning the flows. We are now moving up a level on Maslow's pyramid to social contact and appreciation. We try to make a sustainable lifestyle easy for the people we design for. In our buildings, people like to be together, enjoy themselves, make social contacts, have an outlet for their creativity, and they want to mean something to others and are looking for fulfilment in their lives. That is why we are wnow applying the 7 flows × 7 steps to make our lives more creative, social and enjoyable. The flows are not separate from each other; they are all interconnected. We are therefore going to integrate the flows attractively into our daily lives. We can design them as places where people like to come together. Or places where people become aware of the surrounding nature, the seasons and energy. We are making positive use of our investments in several areas. In the right place, a beautiful tree can also be a sunshade, stimulate biodiversity, enrich your view and provide a place for children to build an epic tree house.

A sustainable lifestyle is not a brake but instead a liberation. People who sail know the magic moment when the engine is turned off and you start to sail. The noise and the diesel stench subside and you are alone with the wind and the waves. You feel the forces of nature, you read them, react to them and enjoy them. With the motor, you are only going from A to B; with sailing, it's all about the journey. You are suddenly switched on in a completely different way. At one with nature. You see the gusts of wind coming over the water, feel the water splashing, the birds flying with you. And you play with your surroundings. The same needs to happen with our buildings, villages, towns and cities. We must accept the natural flows and work with them instead of against them. It's about the journey, the life, not about making money, acquiring status or predictability and the ability to plan. We can rediscover nature and enrich our culture with new customs, new standards and new values. We are building a culture for creative, circular communities. Not going for less bad, just doing things well!

In this chapter, we look at how we can encourage a sustainable lifestyle by designing our flows attractively in the major elements of urban life: living; working and learning; nature and resting points; sport, culture and leisure.

The toolkit for this chapter contains the tool '5 epic moments'. Use this with your team to come up with five epic moments in your building where the users come in contact with the seven flows in an inspiring, surprising or enjoyable way. Moments that will make the future users very happy. A tree house on the roof, a waterfall in a Zen garden, a hanging garden in the entrance hall, edible interior walls or an energy-generating façade. Moments or places that everyone is more than happy to go the extra mile to realize. Make a sketch together or find a picture of what you want to achieve. Briefly describe what it means to you. Later in the process, the architect will replace these sketches with artist impressions, making the epic moments increasingly tangible.

JCA Living Lab, Taipei, Taiwan – JC Architecture Photo: Kuo Min Lee

# Living

## Brings You Peace and Harmony with Nature

Cicero wrote: 'If you have a garden and a library, you have everything you need.' I interpret this as a statement about the private domain, your home where you come to find peace. Where you reflect. It is so wonderfully simple: garden and books. In this quote, inside and outside run seamlessly into each other. Both have endless depth. You can dedicate a lifetime to understanding the garden, making it more beautiful, watching it grow. However, I do miss the other people around us in the quote. Loneliness is a great enemy of happiness, especially in the cities. So for any residential building, the challenge is to combine that simple, tranquil private domain with a clean environment that encourages enjoying a close-knit social community.

I myself lived for a long time in a house in Almelo with a winter garden overlooking a garden with a pond. A very simple room with single glass, open joints between the glass and sliding doors with uninsulated glass. No heating, no ventilation. Just a lamp. And yet in February, with a little sunshine, it was already a great place to sit. And you do sit there, because it is the most pleasant room in the house. The air is fresher, and you can hear the birds, the wind and the trees. It feels much better than sitting inside a house. In February, when the sun goes away, you look at the clouds and gauge whether you can stay put. If the clouds remain, you put on a sweater or just go inside. In spring and autumn, the doors are often half closed and you sit inside in a T-shirt, while everyone outside is still wearing a coat.

In summer, you open all the doors day and night. A dynamic is created between you, your activities, nature and the space.

You find the spot where you like to sit and you control your environment with simple robust means (glass doors open or closed, sweater on or off). You are not going to seal off your entire environment and constantly heat, cool and ventilate to keep everything at exactly 20 degrees Celsius. On the contrary, you are open and adaptable and you accept much more dynamism. You even enjoy it. After a while, you know what time the sun passes the tree, and what time the blackbirds start singing. You see ducks mating in the pond and a squirrel darting through the trees. It is richer and finer and you would so much rather be sitting there reading, eating or talking to others than sitting inside watching television. It is like being able to sail at home instead of going by motorboat.

The new generation of positive-footprint residential buildings makes it easier for a group of people to live sustainably. We start at the scale of the neighbourhood, the residents you may all know by sight. About 50 homes. At that scale, social cohesion can be influenced by means of design, and loneliness and anonymity can be reduced. For this purpose, we arrange natural places where people can meet. With smart green design, we can ensure privacy between the houses, but also create communal meeting places. Choose local plant combinations that change significantly over the seasons, smell good and provide food for birds and insects. Within the neighbourhood, we create a number of places where everyone meets. Everyone comes to the same place. Shared cars, bicycles or letter boxes are concentrated there. There is a seating area if residents want to have a chat, but we also make the 'stuff library' here. Here the neighbourhood shares stuff that you rarely use. Drills, sanders, a gourmet set, a beamer, a basketball. It saves everyone a lot of space and money and is also good for the environment. Energy can also be exchanged through a shared thermal storage facility or a smart grid in the neighbourhood.

In neighbourhoods or large apartment buildings, the residual flows go to the recycling station (there is no such thing as waste!). The design is attractive and clean with plenty of daylight, plants and art. You can think of many other shared spaces that could be valuable in a neighbourhood. For instance, an extra guest room, a DIY space, a gym or a communal kitchen.

In any case, there is a communal garden. Here there is a rainwater tap to fetch water for plants or for a water fight. In the garden, you will also find the compost heap. If there is enough space, climbing areas, yoga spaces or a vegetable garden can be created.

The houses themselves are built in a healthy way; natural ventilation and healthy materials are the norm. You breathe in that air all night, so of course it has to be healthy. On your balcony or in the garden, rainwater is collected to water your plants. In the homes of the future, we must soften the boundary between inside and outside and make it more dynamic. In this way, we can bring nature inside and live our lives more outside.

Every residential building is an opportunity to bring people in sync with their surroundings, nature, the seasons and the natural flows. To give people a place to recharge. Where they feel safe and enjoy themselves in a community to which they contribute. In short, a living environment that gives energy and makes a positive contribution.

A Thousand Yards Botanical Pavilion
Penda architecture & design

# Working and Learning

## The Most Important Daily Meeting Places

When I went diving on a coral reef in Egypt for the first time in 1995, I could hardly believe my eyes. It was so incredibly beautiful! Surrounded by a bare sandy seabed, billions of tiny animals and plants have built something amazing. They are all interdependent, but together they organize their environment based on the natural flows of water, light, oxygen, plankton and the seabed. They are constantly adapting, and creating something together that is increasingly more beautiful, exuberant, colourful and diverse. The diversity of species is immense, with thousands of creatures all having their own place, role, contribution, nursery, shelter and day or night rhythm. They all prefer this bustle to the emptiness of the sea around them.
Our schools and workplaces could be the same kind of meeting place. But we still build our schools and offices based on the idea of an 'average person' and predictability of use. There is limited diversity in places people can work. Spaces are largely the same. We can adjust the temperature three degrees higher or lower, and adjust the sun shading a little. In order to exert maximum control over our environment, we are increasingly filling our buildings with all kinds of technology, apps and control mechanisms. The spaces have to comply with a huge number of rules and regulations that prevent us from creating diversity. Often windows cannot be opened, there is no outdoor space and no one can really contribute to their environment, because everything has to be completely clean again the next morning. Can we also leave freedom and diversity at the heart of our working and learning environments?

The new generation of schools and offices with a positive footprint could be like coral reefs. With a diversity of places where everyone can find the spot that suits them. Where people work inside and outside, open to nature. It works very well if you search with a work group to find the special moments you want to create in the building. Moments that are in line with the mission. Use the toolkit for this purpose. At an insect farm, you make a façade especially for insects. In a primary school, you teach children from the youngest age that waste does not exist by visibly reusing everything. In an accounting office, you jointly come up with something smart relating to visible accounting of the residual flows. You can always devise something to make people enthusiastic and that everyone in the organization can contribute to.

Schools and workplaces can contribute positively to life in many ways. It starts with a healthy working environment, with plenty of daylight, possibilities for natural ventilation, outdoor areas and plenty of greenery. Indoors and outdoors can blend into each other; meetings can also be held in the garden or in a greenhouse. Create natural meeting places in your design by positioning attractive facilities at intersections of your access route.

The materials you choose are also important for health, so no materials, glues or paints that emit toxic vapours, but instead ones that purify the air. Exercise can be encouraged by a tour of the gardens (potentially on the roof), a communal gym or a sports field. In the interior, the recycling areas and coffee points are made into attractive meeting places. There are very few printers, and something only comes out after approval is given by the printer. With the standard settings of computers, energy consumption can be reduced considerably. And let everyone know about energy consumption and generation on a screen. Have the office cleaned during the day during working hours. People are much tidier and better behaved when they look the cleaners in the eye.

You can define some epic moments with the users. A greenhouse on the roof that is a canteen, where vegetables are grown and where the incoming air is pre-heated in winter. An air-purifying pond that is simultaneously a meeting place and a rainwater buffer. Ask those involved in the design process to help find recycled materials and furniture, so that everyone can contribute and has a personal story to tell about the new building.

Community green projects such as an insect garden or vegetable garden generate social interaction and pride. For example, you can visibly collect rainwater via a waterfall and use it to irrigate green walls, which in turn can help with the acoustics. Compensation for trees, biodiversity and soil can also be part of this. For instance, planting trees together in the company wood.

Many of these strategies start with the purchasing department. Their sustainable procurement is crucial for a school or work environment with a positive footprint. So make them part of the design process. As mentioned above, there are many ways to make your workplace or learning environment contribute positively. Each organization can give its own touch here, as long as you make a positive contribution in all the flows. Bringing nature indoors, and bringing out your work and your impact proudly and positively. Together, you can build your own exuberant, incredible coral reef.

*Note: 15 years later, I returned to the same coral reef. All the colour had disappeared, it was as pale as a ghost. Humans have destroyed a thousand years of work on the coral in 15 years.*

Gardens by the Bay, Singapore From 'Glasshouse Greenhouse, Haarkon, 2018

# Nature and Resting Points

## Essential Infrastructure

All over the world, there are city children growing up who will never feel what it is like to build a tree house. Who never fall into the nettles and see caterpillars on the leaves. Who can only climb a tree on holiday. Who have never seen cows and butterflies in real life. Is that bad? Yes, it is. You are a richer person when you understand how nature works, which you are a part of. The immense complexity and coherence of all the flows of water, air, and nutrients and the connection between animals, plants and seasons is endlessly fascinating. But also crucial for us to save the world.

In every construction project, we can take another step towards restoring our relationship with nature. By creating green resting points where we make plants and animals part of our design task. At the moment, we have a kind of 'LAT' relationship (Living Apart Together) with nature. In the last 200 years, since the industrial revolution, we have become detached from it. Because nature is difficult; wild animals are scary. As humans, we crave stability and predictability, whereas nature is frivolous, dynamic and unpredictable. We almost wipe nature out of our cities and allow it only a little in the form of parks, public gardens and trees on the pavement. Nature is okay, as long as we control it. But that's not how it works in a relationship! Certainly not when we ultimately depend on it for our air, food, water and enjoyment. So it's time for us to really start living together with nature. In our homes, cities and on a global scale. Because before you know it, we will not understand each other anymore and the relationship will be over.

When I was studying in London in 2006, there was breaking news: a whale swimming in the Thames in the heart of the city! The whole city was hysterically happy, how incredible! This animal had to be saved. There were helicopters and rescue workers, flowers were thrown into the Thames, everything possible was done. For a moment, wild nature penetrated into the middle of the city. The whale did not survive; its bones are now on display in a museum. The following days, the newspapers were full of reflections on our detached relationship with nature. A week later, London had forgotten. The system keeps on going relentlessly.

When designing and defining your project, try to make room for emptiness, for peace, for real nature. Psychological research shows that nature has a calming and stress-relieving effect on people. Really give plants and animals space. In the city by means of generous parks and tiny forests. Let them go wild by maintaining them less.

It is important to make nature and resting points accessible to everyone, not only in large parks but in every building project. That is why they should not be the balancing items in cost budgets. They should instead be treated as essential infrastructure. Just like water, roads, fibre optics, sewage and electricity. And with corresponding budgets. Designating these items as essential infrastructure helps avoid questions about 'return-on-investment periods'. We don't ask that about roads or sewers either. It should just be there.

In the streets, we soften the surroundings with façade planting and fruit trees. Replace parking spaces with flowers and greenery. Make your buildings green in the façades, roofs and outdoor areas. Programme activities here as well. Make outdoor meetings, teaching or phoning possible. Or just lying in a hammock. Let façades open generously and give everyone a view of greenery. Be lazy too! Let nature take its course; mow and control less. Try to link it directly to the people you are building for. Let them engage in the relationship and take care of animals and plants: put goats on your roof, for example!

What are the benefits of greenery as essential infrastructure? We are making our cities more liveable because they become cooler, less windy and, above all, more beautiful. The colours change with the seasons. Birds sing, butterflies flutter about, and children can build huts. We become healthier, have fresher air and less waterlogging. It smells less dirty in the city and we build a closer relationship with nature that brings us stress relief, inspiration and wisdom.

Île de Nantes, Nantes, Frankrijk Photo: Adeline Gressin

# Sport, Culture and Leisure

## Places for Mental and Physical Health

You walk down a street, and people are dancing and singing everywhere. They ask bystanders to join them, while others watch from a distance with a drink at a pavement café. Some people jog past and further on a group is working up a sweat with colourful balls and ropes. Then behind you someone starts to play the violin beautifully; you hum along softly and enjoy. Is this some kind of hippy dream? Or is it just a description of Covent Garden in London? Maybe both. But this is how a city can be if we dare to let sport, culture and leisure take place outside in the streets again. If we crawl out of the fitness centres, closed cultural bastions and cafés and restaurants and take back the streets from cars and commerce.

We have spoken before of the disruptive division in our thinking between nature and culture. We place ourselves outside nature, as if we were separate from it. When the weather is good, many people go running indoors on a treadmill instead of in the park. Cultural activities such as painting and theatre and museum visits also take place mainly indoors. Sport, culture and leisure can look after our body and mind, like nature. Together they may be even stronger. And wouldn't it be great for all of us if sport, culture and greenery dominated the public space instead of the shopping malls, advertisements and traffic that are so bad for our physical and mental health? Sporting and cultural activities often take place in large buildings that are only open to a limited extent. But when they are open, interesting interactions take place. Can we reinvent this kind of building?

For example, we can make them positive and relevant to the surroundings in their ecological footprint. During the time they are closed, their façades and roofs can provide energy, and purify water and air for the surrounding buildings. Or they can serve as an energy/water/air buffer for adjacent buildings. The roofs could also become green gardens where people can play sports or perform. New connections can also be made: a swimming pool can be heated with the residual heat from a supermarket. The large air-conditioning capacity of theatres can also serve other buildings (or purify city air) during the day. The often-empty car parks around sports complexes should really generate energy or at least be green. Inside the buildings, the materials and air are carefully purified; this is especially important in sports buildings, where people inhale deeply. It is also important to shower economically and to recover the heat from shower water.

When you are in the new generation of sports and cultural buildings, you know that the energy is clean. That the exhibitions are built from sustainable, recycled materials. That the thousands of supporters with their beer and vegetarian snacks in the stadium are fertilizing the fields further away. With every round of beer you also plant a tree. Maybe there will be a big Feyenoord wood near Rotterdam. Because of the often short, intense use of, for example, stadiums and theatres, step 5, 'store what is needed', is particularly important. This allows you to ensure that you have quiet time to purify afterwards, when the building is being used less intensively.

I know that cultural and sports institutions try very hard to reach out to society with the limited resources they have. But it can and perhaps should be more radical. Simply turn everything inside out! The buildings should be opened up, and the activities held outside as much as possible. The streets as the sporting and cultural playgrounds of the city, combined with plenty of healthy nature. Healthy for body and mind. Let these places and buildings in particular be a symbol of the life we want to have – healthy, and dancing and singing through the streets.

III

# How Can I Build with a Positive Footprint?

## Costs, Hidden Costs and Revenues.

## Taking Compensation Seriously.

## It Happens at the Design Team Table!

## Your Drive, Your Dream, Your Role.

# Costs, Hidden Costs and Revenues

## Market Conformity is the Source of Many Problems

Yes, it costs more money. Or to be more precise: a building with a positive footprint requires a higher investment per square metre than a market-standard building that leaves a global trail of destruction in its wake. Or even more precisely: if the trail of destruction were also included in the price of the market-conforming building, the 'cost' would probably be the same. Or, to be entirely honest, it is absurd that we have to provide financial proof for just doing something right. It should be the other way around: someone should be questioned critically why they think they have the right to cause so much damage worldwide with a cheap unsustainable building!

Market conformity does more harm than good. The way investors, authorities, developers, purchasing departments and suppliers hold each other hostage with the concept of 'market conformity' is really too bizarre for words.

Market conformity is linked to standards and prices. But the standards are too low to save the world (and have been lobbied down hard by the industry). As a result, products that fall within the standards can still be full of poison. The 'market-based' prices are too low to make something good. 'But people buy it, so it must be good.' Asbestos was once in line with market practice. Diesel cars are still market-acceptable. 2.3-metre-high storeys once conformed to the market. Similarly, our current standards will be laughable when we look back in 2070. But most of those involved are trying to just meet the minimum standards at the lowest possible cost. This attitude kills everything. That's not how you treat your food, and it's not how you want the school to treat your children.

This attitude also leads to a very big game of hide-and-seek in the construction world. Everyone hides behind standards or market conformity. Because the market is sacred. As soon as construction parties have to do something new or more expensive, the questions follow: Why is that? It's within the norms of the authorities, isn't it? It's not profitable otherwise, is it? I can't just agree to a non-proven technology. This is how many buildings end up being constructed for the minimum price, only just within the standards. Beware of this dynamic at the design team table. As the client, intervene immediately. For the world needs clients and suppliers who do not conform to this perverse market. Who decide for themselves whether they think the standards of the authorities are good enough for their own conscience or for their employees. They build with 'a positive footprint' as the hard non-negotiable lower limit.

### Dealing with Construction Costs

The price per square metre of a building with a positive footprint is currently about 25% higher. How do you solve this? Be radically honest from the start.

1 > You ensure that your budget/m$^2$ goes up by 25%.

2 > You make sure you build 20% fewer m$^2$.

3 > As government, you can depreciate 25% longer. Or as an investor, settle for a return on investment that is 25% lower.

4 > You make it possible with subsidies, tax schemes, total-cost-of-ownership calculations, lease contracts and complicated financial constructions.

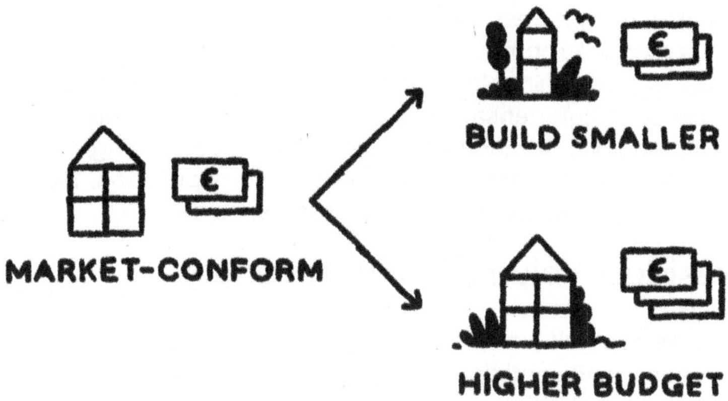


## Size Matters

An interesting option for achieving more budget per square metre is to build fewer square metres for the same budget. Often this can be easily done (except for schools, which we are already building far too small). In programmes of requirements, buffers are often stacked on top of each other, making buildings larger than necessary. Clever layout and double usage make it possible to build much more compactly and still guarantee the desired use. Including the outdoor space and roofs as functional spaces helps as well. With less building, you also use less material and energy per person. As a result, the task of achieving a positive footprint with the available budget becomes easier. In short: size matters. But something also needs to be done in the construction budget to make it fairer and to support the right decisions. You have to add compensation costs and residual value.

## Compensation Costs

Compensation is crucial for the sometimes still unavoidable $CO_2$ emissions and other toxic emissions caused by the construction process and production (more on this later). It costs money to compensate for these hidden costs. If you make more sustainable choices, build less or in a smarter way, you need to compensate less. This makes the environmental costs tangible. Compensation can be done by planting woods and other projects that heal nature or restore pollution. Try to do it precisely where the damage is done.

## Residual Value of Raw Materials

The residual value of the building materials represents future revenues. If you anticipate this, it allows you to incur more costs now. In 2022, experts say this figure will be somewhere between 5% and 10% of construction costs. This will increase in the coming decades due to the scarcity of raw materials and the expected taxation on the extraction of new, non-renewable raw materials. Think of it as a deposit! The rule here is: the better, more sustainable and more demountable your design is, the higher the residual value.

The building cost estimate as we know it remains the basis. But if the benefits of sustainable choices are not directly reflected in the cost estimate, the real costs remain hidden. Your building costs are interconnected with the residual value and compensation costs.

So make sure you put two extra lines into your building cost estimate early on in the process. You can start with the percentages below, which you can refine and optimize during the design process. In this way, construction costs, compensation costs and residual value are all an integral part of the decisions made at the design team table. 10% compensation costs is solid, but it is likely to be less. The 5% residual value is on the low side, this will probably become more. The construction costs for a highly sustainable building are about 15% higher. Together with the compensation costs, this adds up to the aforementioned 25% extra cost per square

**Building Costs of a Positive Footprint Building**

| | |
|---|---|
| Building costs of a market-conform building | 100% |
| Renewable clean flows | + 15% |
| Compensation of indirect damage | + 10% |
| Residual value of raw materials | - 5% |
| **Total** | **120%** |

metre. The residual value already makes up for some of this. This way, you have a fair, robust cost estimate to start with.

A tangible example: if you opt for a wooden construction, you need to compensate less than with concrete. If you choose recycled materials, you need to compensate less than with new products. In this case, your construction costs often go up, but your compensation costs go down.

What if you opt for high-quality modular demountable internal walls instead of sand-lime brick or metal stud? Then your construction costs will go up, but also your residual value (which is shown as a minus at the bottom of your budget). These interventions are crucial. This way, you avoid the cost discussions becoming a 'race to the bottom' in terms of sustainability.

## In Summary

To achieve a positive footprint with your building, the financial aspect is crucial. It must be honest. In addition to the construction costs, you therefore add hidden costs (due to compensation) and future revenues (through residual value). For that matter, it is important to always look beyond the financial returns. This is because the most important benefits cannot be expressed in monetary terms: clean air, water and soil, energy and materials. A pleasant, healthy place for people, animals and plants, and a clean conscience.

# Taking Compensation Seriously

Compensating for your negative effects on the environment is – often rightly – viewed very critically. The unfortunate consequence is that we usually do nothing about it now and just let the damage happen. I see that as a major mistake. In fact, almost all the construction activities we undertake still have a direct or indirect negative impact on several areas: $CO_2$, nitrogen, water pollution, waste, and toxic substances.

We must therefore take compensation for our negative impact much more seriously and embrace it in the design process. A bad product then costs more, because you have to compensate more. For a good sustainably produced product, you need to compensate less, so it becomes more attractive. This way, design considerations become more explicit, reuse becomes more interesting and non-sustainable choices have a direct financial impact as a result of increased compensation costs. Taking compensation seriously also raises awareness of the hidden negative impact that buildings have.

I do agree with the critics that compensation projects are susceptible to cheating, like with that shady $CO_2$-certificate trade. I also agree with them that compensation should only be done when prevention of damage is no longer possible. You also have to be very careful in how you compensate. You will have to find a way to really compensate: on time and not only the $CO_2$ emissions, but also aspects such as any water pollution, eutrophication, depletion and toxicity you cause.

## Environmental Cost Indicator (ECI) as the Basis for Compensation

The compensation method for the new generation of buildings that we are now using is based on the ECI. Based on life cycle analyses, this provides insight into which materials and building components are associated with the highest environmental costs. You calculate the ECI early on during the preliminary design phase. With these insights, you look everywhere for the best possible products and design solutions. The focus then shifts to the five biggest polluters in the search for alternatives, optimizations and better suppliers. Of course, you can do this with all the materials, but often you already know the best application for many materials and there is nothing to optimize.

After each phase, the compensation costs in the budget become more precise. The 'environmental costs' do not directly roll out of the ECI as standard, but are based on the actual compensation projects. When calculating the environmental cost indicator, an overview of the negative impact on eleven impact categories appears. For each category, we look for tangible compensation projects that can offset the negative impact. In your ECI, you can also see directly how favourable the effect of smaller or lighter construction is, which is why you have to keep emphasizing this in the design process. Water-treatment projects, nitrogen-mitigation projects, projects to prevent desertification and $CO_2$-compensation projects by means of planting woods. This method is still in an experimental phase but looks promising.

## Impact Categories

Toolkit 6

1. Depletion of abiotic resources
2. Depletion of fossil-energy sources
3. Climate change
4. Degradation ozone layer
5. Smog formation
6. Acidification of soil and water
7. Eutrophication
8. Toxicity for humans
9. Toxicity for freshwater organisms
10. Toxicity for saltwater organisms
11. Toxicity for terrestrial organisms

# Compensate Generously and on Time

It is important for the balance that you compensate on time and to a greater extent than the damage you do. You must therefore take a wide margin and compensate, for example, 20% more than you calculated. Additionally, the compensation projects must already be in place before you build, otherwise there will be a net negative impact. So, for example, to offset your $CO_2$ you have to plant trees now when you are building next year. It is also good to talk to all suppliers to demand that they improve their production processes, prevent harmful substances and/or eliminate them at source. This is always better than repairing afterwards.

Also, try to communicate about the compensation projects with your users and stakeholders. 'Thanks to us, 10,000 kg of plastic is being fished out of the sea. We have planted 50,000 trees. In Kenya, we have made two square kilometres of desert green and alive.' If they feel involved, it is much more valuable.

The special thing about building with a positive footprint is this: after the building is completed, it will continue to purify. It will further purify the air, purify water and generate energy. The biodiversity and the soil will become richer. The materials will stay clean and the people healthy.

# It Happens at the Design Team Table!

The new generation of buildings with a positive footprint still has to be designed. For this to work well, it is crucial to know roughly what to expect during the design and construction process. The timing is also important. We often discovered during the design process that we should have organized matters earlier. For example, an ecologist usually needs a year to make a proper baseline measurement at a site. In every season, there are different plants and animals to be found at a location. If you call in the ecologist too late and they have less than a year, they cannot do a proper baseline measurement. Below, I share the lessons I have learned with my team, through trial and error. I describe it concisely and chronologically for each phase you will go through. With a checklist for each phase, so you can easily keep track.

## 1 › Select Your Team Carefully: Everyone is Needed

Make sure you gather around you people who have experience with highly sustainable construction and who are willing to go the extra mile. The team should at least consist of the client, architect, structural engineer, installation consultant and building physicist. Ideally, you should also have users, a landscape designer and an ecologist at the table. Let them all read this book first or organize a lecture so that everyone has the same starting point. After that, everyone is more or less on the same page. You can also ask an experienced expert who has already been through a similar design process to give the presentation. At a first meeting, give a personal account of your own drive, dream and role for this building. Ask them to do the same. By doing this, everyone gives their personal commitment. Then, in one or two sentences, define the team's mission for this positive-footprint building. Use the Toolkit for this purpose.

It may be that the subject matter is still rather abstract for some people, in which case visiting a radically sustainable building works very well. The city council offices in Venlo or the Triodos Bank at De Reehorst country estate give people energy and make the value of a positive footprint tangible (although these buildings are not that yet in every respect!). You can all collect some best practices taken from the literature as well and present them to each other.

- > Select a capable positive-footprint team
- > Have everyone read this book
- > Hold a personal-commitment session Toolkit 2
- > Define a team mission Toolkit 2
- > Share best practices in short presentations

## 2 › Definition Phase

Formulate together, on a single A4, the specific ambitions in terms of positive footprint for your building project. These can be general ambitions such as: 'We want to purify air.' But they can also be more specific: 'We want to purify the motorway air into forest air. We want at least 100 plant and animal species at the location after completion. We are going to reuse 100% of the materials from the existing building on the site.' These ambitions are your compass during the design, construction and life cycle of the building. This is what you are fighting for, and everyone should be able to envision it in their sleep. The most important thing is that the client and users express their commitment and feel attracted to the ambitions. That is where the success begins, and they will demand it afterwards from the engineers and designers.

As part of the definition phase, it is also important to critically examine the programme of requirements. Size matters most at this stage. Can the same use be achieved in a smaller building for the same budget? The residual value and compensation costs are added to the building cost estimate or the budget.

> Commitment of client to ambitions
> Commitment of users to ambitions
> Build-smaller check of the programme of requirements
> Compensation costs and residual value in the budget

## 3 › Baseline Measurement

In the baseline-measurement phase, you determine what you approximately need from the various flows and what is available at the construction site. Provide a baseline measurement immediately of the exact quality and quantity for the seven flows. It means that you know your starting point for the quality of water, air, soil, energy and biodiversity. You have checked whether there are residual flows for building materials in the immediate vicinity and have a feel for the products that are needed during use.

You have determined how much water you approximately need and will have to purify. You have looked at how much rainwater you can collect, how much energy you approximately need and how much you can generate from the sun and so on. Use the baseline measurement tool in the toolkit.

You also quickly see where you can have a major positive impact and where it becomes difficult. This step often takes longer than you had anticipated; the ecologist, in particular, needs time. So start this process in good time. The purchasing department, if any, can be involved at this stage. They often know best how much energy, paper, coffee and water are used. For example, on the basis of purchase receipts from a whole year. These figures are a good start for getting a handle on the quantities. Later, they can help make the flows more sustainable. The building services consultant and building physicist can estimate the amount of water, air and energy to be used and what is available in the surrounding area.

> Ecologist first at the table for baseline measurement
> Baseline measurement and quantification of required flows Toolkit 3
> Analysis of flows in vicinity
> Purchasing department involved

## 4 › Generate 7 × 7 Flows Design Sketch

Once the ambitions, frameworks and flows are clear, you can start designing. You design for people, animals and plants. But don't just design your building and the interior. Include all the flows in the design as well, and make it a regular agenda item at your design team meetings. Make sure you get a handle on the quantities and the space that facilities need. You fill in the matrix in draft form together: what are we going to do to store energy? How are we going to purify the air? Discuss different scenarios for the seven flows. Draw the facilities that need flows early on, so that they become visible to everyone. Sometimes they turn out to be too big and you have to find another solution. Consider, for example, the space that a helophyte filter takes up to purify the water. This is usually not possible in the city because there is not enough space. You can see that quickly and then you know that, for example, you have to investigate a bioreactor and give it a place.

Now that the building is beginning to take shape, it's a good idea to devise the five epic moments together where flows can be experienced in an attractive way for the users. Use the toolkit for this purpose. Also check once again with the client and users to see if they can come up with clever ways of sharing space and building smaller. You can also explore with each other how to use fewer large flows in the future.

> › Put positive footprint on the agenda of each meeting (also steering committee)
> › Generate scenarios for clean flows
> › Devise five epic moments with users Toolkit 5
> › Complete draft matrix jointly 50% Toolkit 4
> › Build smaller through smarter use of scenarios

## 5 › Preliminary Design

Working from the flows, brainstorm how to connect them so that they add value to each other. A green roof with a lot of biodiversity can also purify the air. Purified grey water can be used to irrigate plants, etc. Design the flows so that they become visible and can be experienced, to give the end users more pleasure, health and inspiration. Check at every meeting: is everything still positive? Are there any new ideas or techniques? During the preliminary design, you fill in the matrix more precisely. Some solutions are already becoming apparent. At the end of this phase, you make an ECI calculation. You can see what impact your building will have on the environment at that time. Based on this first analysis, you will optimize your design in the next phase and look for relevant compensation projects. It is also wise to create a few appealing images and diagrams in the preliminary design that show everyone involved how the flows will run, and how attractively they integrate with everyday life in the new building. Update the five epic moments in the toolkit with impressions. This can be used to rekindle the spark, so that the people involved see what they are working so hard for. The users are now really looking forward to their new building, as it is becoming more concrete.

> Put positive footprint on the agenda of each meeting (also steering committee)
> Design the clean flows
> Update the five epic moments Toolkit 5
> Update the matrix jointly 70% Toolkit 4
> Calculate ECI and compensation costs Toolkit 6

## 6 › Final Design and Permitting

You encounter resistance, or things are more expensive or more difficult than expected. In this phase, the difference is made between good intentions and an actual positive footprint. Dig in your heels and refuse to let go of the ambitions. Do not point the finger at each other, but try to find smarter solutions together. The user involvement that you created in the preliminary design comes in very useful. You optimize and design the flows to be attractive and robust. You choose your materials carefully and make sure they can be demounted. You fill in the matrix again with the latest design and calculate the ECI. You use these for a last optimization exercise before finalizing the specifications. You also use this to fill in the compensation tool in the toolkit, and look at what type of projects you want to select to compensate.

- Put positive footprint on the agenda of each meeting
- Design the clean flows to be robust and attractive
- Update the matrix jointly 90% Toolkit 4
- ECI calculation optimize design Toolkit 6
- Compensation Toolkit 6

## 7 › Technical Design

Present everything at each phase to the users and to as many stakeholders as possible. It gives them energy as well and many ideas come your way. Commit key suppliers now. When you make your specifications, you actually want to be in dialogue with them, so that they too can innovate. You detail and specify the building design with the information from the suppliers you have selected. This allows you to create the draft materials passports, the final ECI calculation and demolition specifications. Your objective is to know what is in the building and how you can get it out again in a number of years with maximum residual value. Once the ECI calculation is final and you have selected the compensation projects, you start these projects. In the specifications, it is important to set high standards for all products, production methods and the transport of all products.

- Put positive footprint on the agenda of each meeting
- Create materials passport + demolition specifications
- Detail the clean flows with product suppliers
- ECI calculation and final compensation Toolkit 6
- Start compensation projects

# 8 › Construction

The compensation projects are already running, so the world is already getting a little better. During construction, you continue to work on the positive footprint. Together with the users, you start by communicating widely about the positive footprint and what this means for the use and the functioning, and what everyone can contribute to it. Involving everyone often leads to new, useful ideas and initiatives. Make sure the selected builders and suppliers are also aware of how you are trying to achieve a positive footprint. Also ask them how they can contribute to this aim. For instance, with electric transport, preventing waste, reusing cutting losses, reusing residual flows, and product innovations. You can also give the builders a presentation about the objective and the process so far to make them enthusiastic. At the end of construction, the materials passport and demolition specifications are delivered by the builder. These are crucial for the residual value of your building.

> › Put positive footprint on the agenda of each meeting
> › Communicate externally
> › Prepare for sustainable use
> › Deliver the materials passport

# 9 › Building Taken into Use

Every building starts with teething troubles, including a building with innovative techniques and a positive footprint. Be open about it beforehand and prepare everyone for the situation. In the first few months, everyone has to allow the building to settle a little. For critics, the teething problems will be a confirmation that innovations or the positive footprint do not work. Take them seriously but stay the course, and try to remedy teething problems as quickly and transparently as possible. Together with the suppliers. When the building is taken into use, it is also important to help people with their new, sustainable lifestyle. You can organize 'onboarding' for this purpose. Keep spreading the word about the benefits during this process as well. Sustainable use also includes sustainable maintenance. Involve the maintenance parties and make them part of the onboarding. Ultimately everyone is proud, so make it a great celebration! After a few months, you verify if the air, water and energy flows are actually clean and running as predicted. You will have to make adjustments, if necessary, but hopefully after a few months you will be able to proudly communicate how much water and air have already been purified.

> Verify matrix
> Onboarding for sustainable use
> Sustainable maintenance contract concluded

Toolkit 4

## 10 › Life Span

After a year of use, measure whether the positive footprint is actually being realized. If not, then you talk to the design team and the suppliers. Where does the problem lie and how can it be remedied? Also check whether the users are experiencing everything positively. Communicate the results!

> Annual report on the positive footprint
> Monitor biodiversity

## 11 › Updates

The use of a building changes constantly. In the design process, you try to anticipate this and to build in flexibility. Sometimes, however, there are unexpected changes that require an update. If you are updating the building, it is best to involve the consultant or architect originally responsible. Together you look at what is needed. You use the demolition specifications to guarantee maximum residual value when changing or updating, and you update the materials passport.

> - Contact the consultant and supplier originally responsible
> - Check demolition specifications before you start rebuilding
> - Update demolition specifications and materials passport

Designing and building is a complex process that cannot be summed up in a simple checklist. Nevertheless, the above checklist helps to bring some predictability into an exciting adventure. And the toolkit helps to bring your dream of building with a positive footprint closer more quickly.

# Your Drive, Your Dream, Your Role

You have now learned the method 7 × 7 = positive footprint to realize a building with a positive footprint. You know roughly what you will encounter during the design and construction. You know what to look out for. And I have tried to give arguments and examples everywhere that you can use to overcome resistance or win over those involved in the construction process. But what is the very first step? I think it starts with knowing your own drive, and the drive of the people you will be working with. Expressing a personal commitment helps a great deal.

Some time ago, I was in a presentation where I suggested to the client that they make a building with a positive footprint. I told him that it would cost a lot of additional effort and also extra money. Oh, and the outcome was also uncertain, because we still had to innovate substantially in a number of areas. But I also told him what benefits it would bring for his employees and those around him and how much adventure is involved in this process. When I had finished, he said, 'Great story. But why do you want this?'

I thought it was a good question, but a strange one. Why should he care why I want this? The question is whether he wants it! I had thought about it, but I had never formulated it explicitly. I improvised an answer, but did tell it straight from the heart. We were given the assignment.

My answer was rather more improvised, but boiled down to the following:

*The most beautiful thing I have ever seen in my life was a coral reef in Egypt. Fifteen years after I went, everything there was dead!*

*I really wish my daughters could have enjoyed that beauty too. I want to continue designing buildings all my life. But not if they only make the problem worse, so I have to do things differently now [my drive]. That is why I dream of a world full of buildings that actively make their surroundings cleaner [my dream].*

*I am an architect, I can use the power of design to integrate systems and flows in a new, better way. I can also make the future users enthusiastic with a beautiful, rich design where nature and architecture, inside and outside, flow into one another. This will later allow them to work in balance with their natural environment and enjoy it much more! [my role].*

Only later did I really understand how smart and important the client's question was. By letting me answer his question, he suddenly knew me. He understood what I dream about and therefore what I am moving towards. He understood my intrinsic drive and the role I play in realizing that dream. And because he asked for my personal response, and not my offices's, I also implicitly gave a deep personal commitment to do absolutely everything possible to realize this building with a positive footprint.

All in all, a very simple question that strengthens the team dynamic and mission in a very powerful way. Through mutual understanding and personal commitment. So start the design phase by writing down your own drive, dream and role. Then, in the first design team meeting, ask your design team members what drives them personally to create a building with a positive footprint. Every time someone joins you during the design and construction process, ask them as part of the introduction round about their drive, dream and role in realizing a building with a positive footprint. That way you'll gave a committed team and you quickly get to know each other a bit better.

This is your workbook, so you can write down your own drive, dream and role here. Then you won't have to improvise at your next presentation.

# Toolkit

positivefootprint.nl/toolkit

Toolkit 1

# Checklist Phases

## 1 › Select team

- O Select capable positive-footprint team
- O Have everyone read this book
- O Personal commitment session — Toolkit 2
- O Team mission — Toolkit 2
- O Share best practices in short presentations

## 2 › Definition phase

- O Commitment of client to ambitions
- O Commitment of users to ambitions
- O Build-smaller check of programme of requirements
- O Compensation costs and residual value in budget

## 5 › Preliminary design

- O Positive footprint on agenda of each meeting (also steering committee)
- O Design the clean flows — Toolkit 4
- O Update matrix 70% — Toolkit 5
- O Update 5 epic moments — Toolkit 6
- O ECI calculation and compensation costs

## 6 › Final design

- O Positive footprint on agenda of each meeting
- O Design the clean flows to be robust and attractive
- O Update matrix 90% — Toolkit 4
- O ECI calculation and optimize design — Toolkit 6
- O Compensation — Toolkit 6

## 9 › Building taken into use

- O Verify matrix — Toolkit 4
- O Onboarding for sustainable use
- O Conclude sustainable maintenance contract

## 10 › Use

- O Annual reporting on the positive footprint
- O Monitor biodiversity

## 3 > Baseline

- Involve ecologist first for baseline
- Baseline measurement and quantification of required flows Toolkit 3
- Analyze flows in surroundings
- Involve Purchasing

## 4 > Design sketch

- Positive footprint on agenda of each meeting (also steering committee)
- Generate scenarios for clean flows
- Fill in matrix in draft form 50% Toolkit 4
- Devise 5 epic moments with users Toolkit 5
- Build smaller through smarter use

## 7 > Technical design

- Positive footprint on agenda of each meeting
- Create materials passport + make demolition specifications
- Detail the clean flows with product suppliers
- ECI calculation and final compensation Toolkit 6

## 8 > Construction

- Positive footprint on agenda of each meeting
- Communicate
- Prepare for sustainable use
- Deliver materials passport

## 11 > Updates

- Contact consultant and supplier originally responsible
- Check demolition specifications before starting any alterations
- Update demolition specifications and materials passport

positivefootprint.nl/toolkit

# Toolkit 2 Drive, Dream, Role, Mission

## Your drive

**Where does your personal drive come from?**

## Your dream

**What is the dream you are chasing?**

## Team mission

Describe in two sentences the team's positive-footprint mission

## Your role

How do you see your role? And what are you going to do to fulfil it?

Toolkit 3

# Baseline Situation

| | Present at location |
|---|---|
| | Measure the quality of the flow and the amount present, and add the report and amount below. In total or per year. |
| **Air**<br>$m^3$ / year | |
| **Water**<br>Drinking water $m^3$ / year<br>Grey water $m^3$ / year<br>Black water $m^3$ / year | |
| **Energy**<br>kWh / year | |
| **Soil**<br>$m^3$ clean<br>$m^3$ polluted | |
| **Biodiversity**<br>Types of plants<br>Types of animals<br>Specials | |
| **Building Materials**<br>Biobased kg/$m^3$<br>Technical kg/$m^3$ | |
| **Food**<br>Electronic equipment, food, furniture, paper, beakers, coffee etc. (kg/$m^3$) | |

## Required

To get a feel for the scale of the flows, fill in how much of each flow you need below.
In total or per year.

| Toolkit 4 | Matrix Positive Footprint | | |
|---|---|---|---|
| | 1 > Formulate the Positive Ambition | 2 > Measure the Baseline Situation | 3 > Harvest-focused |
| Air | | | |
| Water | | | |
| Energy | | | |
| Soil | | | |
| Biodiversity | | | |
| Building Materials | | | |
| Food | | | |

| 4 › Use and Enjoy Wisely | 5 › Store What is Needed | 6 › Purify Afterwards, Out of Love | 7 › Measure and Communicate |
|---|---|---|---|
| | | | |
| | | | |
| | | | |
| | | | |
| | | | |
| | | | |
| | | | |

positivefootprint.nl/toolkit

Toolkit 5

# Epic Moments

1

2

3

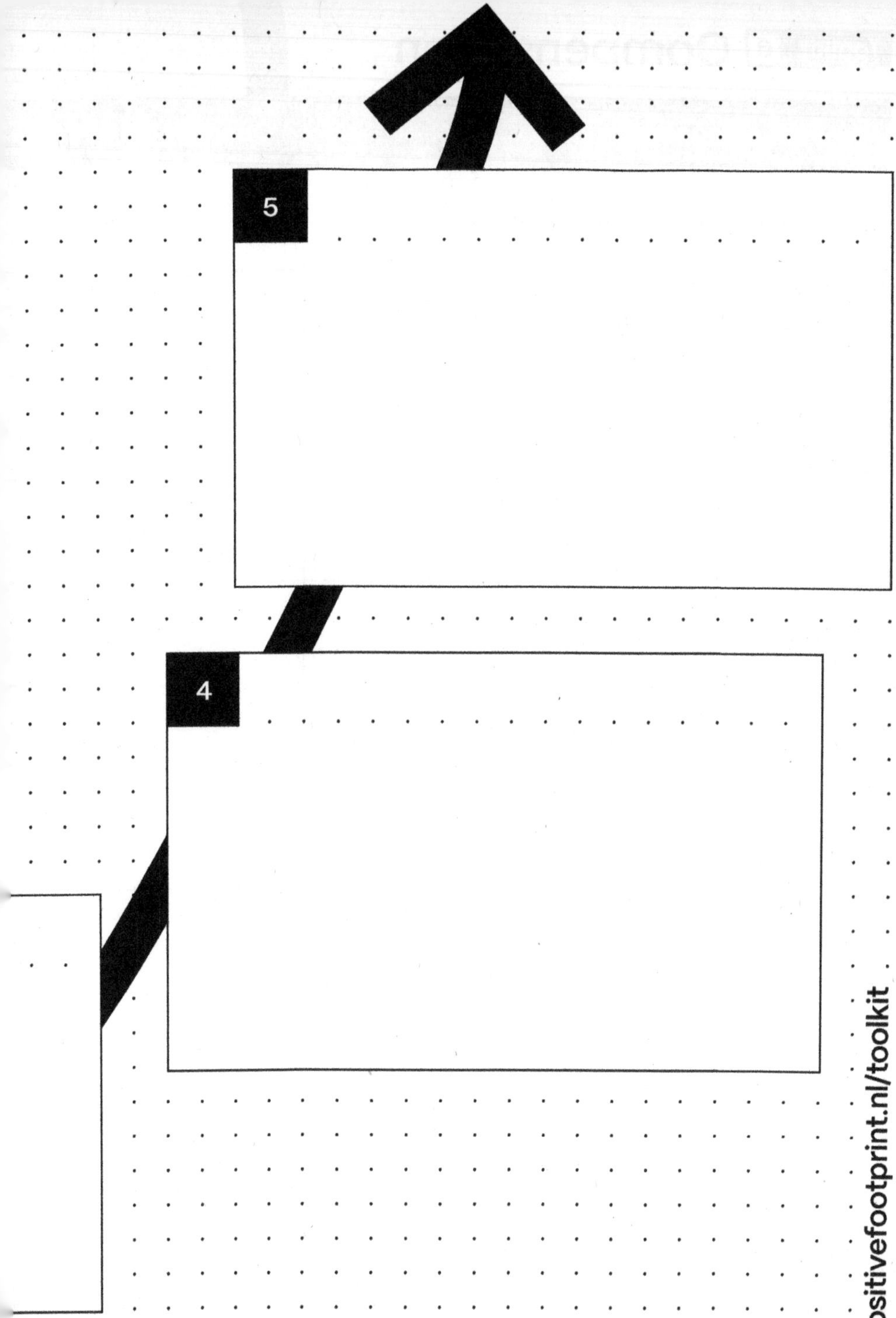


positivefootprint.nl/toolkit

# Toolkit 6 Compensation

| Impact categories | Quantity (from ECI) |
|---|---|
| 1. Depletion of abiotic resources | |
| 2. Depletion of fossil-energy sources | |
| 3. Climate change | |
| 4. Degradation ozone layer | |
| 5. Smog formation | |
| 6. Acidification of soil and water | |
| 7. Eutrophication | |
| 8. Toxicity for humans | |
| 9. Toxicity for freshwater organisms | |
| 10. Toxicity for saltwater organisms | |
| 11. Toxicity for terrestrial organisms | |

## Compensation projects

positivefootprint.nl/toolkit

'If you want to build a ship, don't drum up the men to gather wood, divide the work, and give orders. Instead, teach them to yearn for the vast and endless sea.'
— Antoine de Saint-Exupéry

Epilogue

# And now Build Well!

I have been working as an architect at Kraaijvanger Architects for 15 years now, where I have learned a great deal about sustainable building and living. Also about the construction industry as a system. Four times I have succeeded in getting the design team into a real flow to create a building with a positive footprint. I should say: we succeeded, because you never do it alone. Why did it fail so often? The ambitions were not clear; we were not on the same page in the team; nobody knew exactly where to start or I did not know how to communicate the strategy and necessity clearly enough. In most cases, everyone did want to do all sorts of things, but the client didn't dare in the end. And yet it must be done, always, for every building. There is often not enough time to explain in depth what the paradigm shift means, so I wondered what I could do about it myself.
I have written a guide that is as compact as possible, so that everyone starts from the same point. 'Fewer than two hours' reading, so everyone should be able to manage that.' This way, the design team is quickly on the same page and moving in the same direction. You're in the flow!

While I was writing this book, I was already doing better: I can easily get most clients and design team members enthusiastic. And I am working on several designs that are now pursuing a positive footprint in an inspired way. With more energy, a greater sense of adventure, more inspiration, more difficult puzzles, more creativity and much more satisfaction. And now let's get building!

Use this book for your building projects and give it to the designers and others you work with. Buy one and pass it around, or buy ten copies and hand them out. I am very curious to know if it has helped you. I am now continuing to work with many inspired colleagues and partners on this new generation of buildings. It's going to be an incredible generation! We will share as many experiences as possible on positivefootprint.nl to help others further in their work. If you have a useful tool or a good design that others can learn from, we will post it on positivefootprint.nl. Also, if you have any questions or just want to spar: please contact us!

Contact: vincent@positivefootprint.nl

Download the toolkit: www.positivefootprint.nl

# Thanks and Sources

The dream of realizing buildings with a positive footprint is not just mine alone. That dream has grown together with those of my colleagues at Kraaijvanger Architects. I have been completely entwined with the office for 15 years, and in those years we have learned and innovated an incredible amount. In that time, there are so many people who teach you things, who influence you, question you critically, give you a push in the right direction and inspire you, that I don't know where to start thanking them. So thank you all.

I am the one writing it now on behalf of all of us, but we have let the idea of building with a positive footprint grow together. And we are going to make it happen together. It is wonderful to see this architectural office reinventing itself to design a new generation of buildings. Buildings with a positive footprint.

Bart van der Werff, Mijke Rood, Rinske Wikkerink, and Mauric Cornet are important 'partners in crime' and I would like to thank them personally for their contribution to this book. Their drive, expertise and enthusiasm were indispensable. I would also like to thank my predecessor Herman Kraaijvanger, whom I never knew. His wonderfully concise and very direct book *Hoe zal Rotterdam Bouwen? (How will Rotterdam Build?)* was published in 1946. When I reread it two years ago, it inspired me to write this book.

I would like to thank everyone who has made *Building with a Positive Footprint* possible. First of all, my daughters Elise and Yaëlle for their smart questions, fun and inspiration and my wife Laurence for her patience. Second, Paul Swagerman and Loek Vugs for the book design and Marcel Witvoet at nai010 publishers for publishing. I thank the critical proofreaders, the inspiring partners and the courageous clients who dare to take on the adventure. Special thanks go to the authors of the books that have inspired me and made me wiser. I am particularly indebted to Michael Braungart and Bill McDonough for their insanely inspiring cradle-to-cradle books, and also to *Doughnut Economics* by Kate Raworth and *Net Positive* by Paul Polmans. A list of sources and other interesting literature will be published on www.positivefootprint.nl.

# Colophon

**Texts**
Vincent van der Meulen

**Copy editing**
Els Brinkman

**Translation**
Christine Gardner

**Proofreading**
Leo Reijnen

**Graphic design**
Paul Swagerman

**Illustrations**
Loek Vugs

**Publisher**
Marcel Witvoet
nai010 publishers, Rotterdam

ISBN 978-94-6208-745-3
NUR 648
BISAC ARC009000, ARC018000

*Building with a Positive Footprint* is also available as an e-book:
ISBN 978-94-6208-753-8

*Building with a Positive Footprint* is also available in Dutch:
*Bouwen met een positieve footprint*
ISBN 978-94-6208-744-6



nai010 publishers is an internationally orientated publisher specialized in developing, producing and distributing books in the fields of architecture, urbanism, art and design. www.nai010.com

nai010 books are available internationally at selected bookstores and from the following distribution partners:

North, Central and South America – Artbook | D.A.P., New York, USA, dap@dapinc.com

Rest of the world – Idea Books, Amsterdam, the Netherlands, idea@ideabooks.nl

For general questions, please contact nai010 publishers directly at sales@nai010.com or visit our website www.nai010.com for further information.

Printed and bound in the Netherlands.